RICARDO

50+

IP HACKS

www.50-ip-hacks.com

With a big thank you to my family, EVOip team, partners, and customers who enable this book.

IP ADVICE FOR
IP PROFESSIONALS

THIS IS NOT LEGAL ADVICE – JUST SOME FRIENDLY ADVICE!

This book will NEITHER tell you about the advantages or philosophical aspects of proper IP Management, fine English literature, nor is this book a replacement for any professional legal advice. This book will tell our readers about hands-on hacks, tricks, or ideas to apply to master your IP Management.

Please be advised that each *Hack* is to be seen on its own and must be evaluated by you if it's fit to your contexts and shall be discussed. So, this book is only for IP Professionals or IP- minded people who would like to get other impulses to reflect and act on their IP Management critically.

Mannheim 2022 | EVOip GmbH

 FIRST EDITION.

ISBN: 9798764521916

PREFACE

Hello there,

My name is Ricardo Cali, I am the lead-author of this book and Managing Director of a German IP Engineering company called EVOip. I spent years mastering IP Management and learned a lot from my encounters with people - not textbooks. After several years working with some of the best IP professionals of the corporate world who have influenced and inspired me, I decided to share my learnings over posts on LinkedIn. With this book, we go one step further, and I have the intention to share 50+ of my learnings and thoughts that I have had so far.

This book is a continuous work in progress, and some hacks will be redefined, altered, or replaced by other IP experts in the years to follow. If you are interested in participating, please let me know and contribute with a nice hack.

(Via E-Mail to contribute@50-ip-hacks.com)

Compared to the Martial Arts world, this book is not about *budo*. This book is about straight & dirty self-defence techniques. If you are looking for a bigger picture, motivation, or scientific method, this book will not help you. This writing is intended to be something like *Krav Maga,* easy to understand, time-saving and effective. You won't find all the answers to IP Management in this book, but you will find 50+ ideas, thoughts, and tactics, also known as *hacks, which may help with your IP Management* tasks. If only one of these *hacks* can work for you, your spent time has been worthwhile.

This book is constructed to fill your everyday waiting periods most efficiently. The *hacks* are written in such a way that each can fit in one or two pages, have a short explanation, and if applicable, a resource, and a call to action. The *hacks* are not listed in any specific order or related to each other so that you can start reading anywhere in this book.

Enjoy reading!

Mannheim, 21st February 2022

Ricardo Cali

INNOVATION AND IP CREATION

PATENT STRATEGY

PATENT STRATEGY – The fundamentals

PATENT STRATEGY – Patent budget monitoring

IP MANAGEMENT

ABBREVIATIONS

AI	Artificial Intelligence
ASAP	As soon as possible
C-level	The leadership positions at the top level in the organizational chart of a company
DIN	The German Institute for Standardization
EPC	The European Patent Convention
EPO	The European Patent Office
EU	The European Union
FMEA	Failure mode effect analysis
FTO	Freedom to operate
HoK	House of Knowledge
ID	Invention Disclosure
IDs	Identifications
IP	Intellectual Property
IPM	Intellectual Property Management
IPR	Intellectual Property Rights
ISO	The International Organization for Standardization
IT	Information Technology
KPI	Key Performance Indicator
PCT	The Patent Cooperation Treaty
QM	Quality Management
R&D	Research and Development
TESE (of TRIZ)	Trends of Engineering System Evolution
TRIZ	The Theory of Inventive Problem Solving
UX	User Experience
WIPO	The World Intellectual Property Organization

INNOVATION AND IP CREATION

1. Innovation and IP Management are like gardening.

Focus on removing blockages and limitations.

For me, Innovation and IP Management are like gardening.

No matter how hard you work, the fruit will grow by themselves, at their own pace.

You can only influence them by providing them the correct conditions and removing as much vermin as you can.

The same goes for ideas and inventions. You can only provide everything they need to proceed, and the market will decide whether they are "delicious."

Prosper gardening will help turn the odds in your favour. But ***how?***

In Germany, there is a popular strategy, the "Wolfgang Mewes Bottleneck-focused strategy," which states the following:

A plant will grow to a certain level if it gets enough water, soil, air, and sun. All factors have to be fulfilled, and the plant continues growing until one factor runs out and becomes a bottleneck.

According to Mewes, such bottleneck has to be analysed and solved for business development and market leadership.

So, what is the bottleneck in your Innovation and IP Management?

Resource:

Cybernetic Management Theory (EKS). Mewes Verlag, Frankfurt am Main, ISBN 3-922062-00-8 ..

2. You want to achieve success in innovations?

Invest in a properly staffed, budgeted, and working IP Management.

What your top management needs to learn:

If you want to succeed in innovation, you need a properly staffed, budgeted, and working IP Management.

It won't matter how good your product is if you run into potential infringement issues, because this will turn bad for the organization.

So please take time to choose the right people in YOUR IP team, provide them with enough money and support so that they can do their jobs the best way they can, which will, in turn, make everyone else do their jobs better.

But what is a clever way to do this?

Let them receive first-hand experience. Provide them with case studies, show them examples of your (or others') effective IP Management cases.

Let them figure things out by themselves. Do not provide the answers. Let them answer their own questions like where to place IP in their priorities.

3. IP = Innovation?

What do you think? By Dr. Claas Gerding-Reimers.

Do you cringe when it comes to easy and misleading definitions? Well, I often do. So, a while ago I asked a quite short question: Is IP=Innovation? It fostered more discussion in my network than I expected. It was great to see that many experts put their minds on a quite "basic" issue of definitions and interpretations around "IP", "innovation", "invention" and more. (See the first link below to my summarised blog post. A big thanks to all contributors for the initial discussion.)

Here are the 5 major takeaways:

1. Inventions, IP Rights and Patents can be highly useless, if they are not commercialised (as product, service, license etc.).
2. An innovation without IP is questionable, usually IP is part of the process in creating an invention, which then produces a form of IP.
3. IP can be seen as "an" indicator of innovation. It is neither more than that, nor it is less.
4. Vested interests / misunderstandings are the reasons that patents are reported as THE indicator of innovation: "Most innovative countries/companies hold the biggest quantity of patents or patent applications." No!

5. Taken together, “IP=Innovation” is an easy, shortcoming and misleading equation. Instead, a better, simpler and catchier definition comes from the MIT (see 2nd link below):

Innovation = Invention x Commercialisation

Still, one can find “innovations” that do not fit into this equation, as well. Only a few considerations: What is an invention? Does an invention always have to be “commercialised” to create value and thereby become an innovation? Would “exploitation” be a more appropriate term than “commercialisation”?

IP in this connection is a powerful tool which can provide huge value in leveraging the commercialisation potential of an invention.

So better don’t miss out and get your IP management right to create those valuable innovations.

Resource:

https://ocw.mit.edu/courses/sloan-school-of-management/15-390-new-enterprises-spring-2013/video-tutorials/lecture-2

4. Patent protection validity check.

The design-around question.

The value of your patents depends on the job you want them to get done.

When we think about patents, we think about exclusivity. And the value of your patents is the time it takes to find a work-around to your solutions or to invalidate your patents, like most start-ups wish to.

The truth to be told is however, that patent holders not always consider other available solutions or public domain solutions, which can render their perceived non-existent exclusivity.

Therefore, here are quick fixes to identify how well your patent is protecting you:

- Conduct an infringement analysis of your patent with your own product. If there is a clear overlapping detected, you have a good patent – product coverage.
- Start to design around your patent to make your own product outside the scope of your patent's protection.
- From the work results, you can decide whether to obtain patents for these design arounds or reconsider your patent protection and investment.

Patent protection validity check:

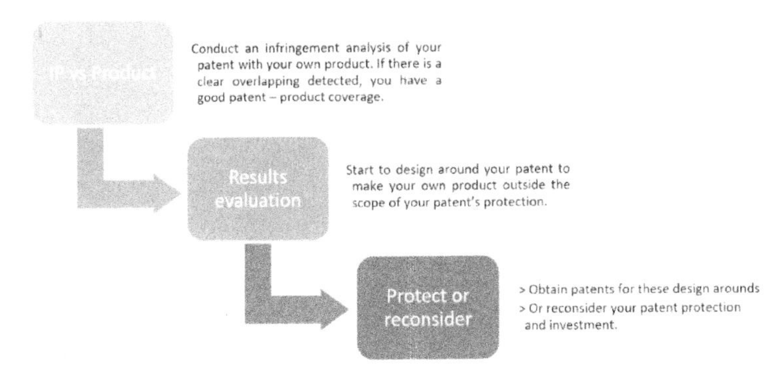

Additional resources for design around's understanding:

https://en.wikipedia.org/wiki/TRIZ

https://en.wikipedia.org/wiki/Patent_infringement

5. Strategic IP Creation to ring fence competitors.

Easy five-step method to ring fence competitors.

Strategic IP Creation to ring-fence competitors is actually producing patent paper tigers with surprisingly sharp claws.

Ringfencing can either be reactive or proactive. In both cases, you need a small team to solve this. But usually, ring-fencing is requested as soon as an incident happens. You need at least one IP-savvy person and one engineer. But all this will take time if you are not prepared.

Therefore, the following is a quick method that you can use for this:

1. Start with a reference (own competitor's product analysis or third-party IP).
2. Analyse potential shortcomings and anticipate failure modes. You can use FMEA (Failure Modes and Effects Analysis) Methodologies.
3. Solve them.
4. Score and select the most promising ones.
5. Create filings that will ruin someone's day within 18 months.

The average time it takes to undertake this effort can be calculated as follows: 2 days of preparation, 4 days of ideation, plus 4 weeks for description (based on our experience with multinationals).

In order to be ready for such a case, do an annual exercise together with selected engineers.

6. Do not only look for a big fish!

Cluster your ideas to build a bigger idea.

If you run ideation sessions or innovation workshops, you might not discover the one breakthrough idea - *the big fish*.

You will get many "little" ideas. The trick is to cluster and develop them. To form a "swarm of ideas" and understand its direction where it flows in terms of user benefits.

These ideas shall overcome the problem to be solved or the technical or physical contraction; and provide the intended benefit for the customer.

You need to combine, merge, and absorb ideas. While doing these actions, you will be able to generate new ideas.

That's why your ideation session shall be divided into two appointments, to give the attendees more time and let them merge ideas in their head.

So, your tiny ideas will become the big fish. And big fishes are much easier to patent.

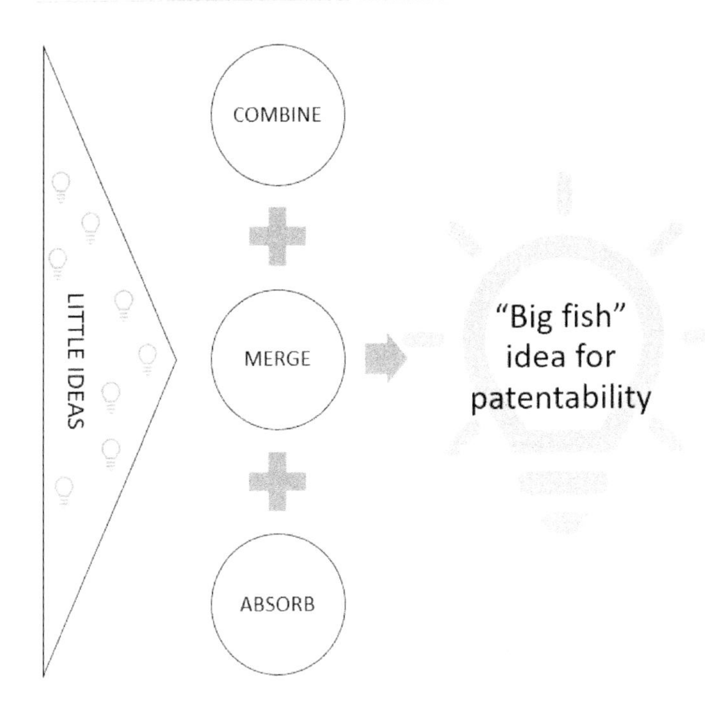
LITTLE IDEAS
COMBINE
MERGE
ABSORB
"Big fish" idea for patentability

7. Tips for more invention disclosures.

Three steps to receive more invention disclosure submissions.

Invention disclosures are usually required in written form. The requested details and amount of information necessary to be accepted as an invention differ from organizations to organizations.

My thesis is all about the design of the invention disclosure experience which has direct influences on the total amount and quality of submissions.

And my approach for more qualitative invention disclosures consists of three steps:

Step 1: Make it as ***easy*** as possible and cut down all unnecessary fields.
As an example, do you really need to know, as an inventor, what prior art is? Or other applicable industries?
My recommendation is to keep only the mandatory fields and add other fields after the submission. Offer phone numbers, support chats, and contact details on the invention submission information in case they ask questions.

Step 2: Make it as ***convenient*** as possible. And offer a full-service experience, ideally with the one-stop-shop approach.

If you are an employee, who is always busy, inventions are not your key responsibility, and the IP department could be delicate to deal with. Do you really think an employee has any desire to submit anything then? Therefore, improve the experience, for example by setting up a kind of "hotline", providing a skilled person to draft and helping with the invention disclosure. Furthermore, assign an engineer or IP expert who can help reduce that idea of the inventor to practice, so that the concept is sufficient.

Step 3: If no one provides it, collect it.

Sometimes people don't have the time to complete or feel that their concept is not complete yet. Usually, the draft is already sufficient for starting the IP generation process.

Therefore, reach out to the departments or employees and ask them if they have any ideas or is there anything you could help them with.

Proactively meet with the Project managers of nearly completed projects to ask if they have an additional matter or previous solution to protect.

If you have an entry hall or anything similar, place a stand in the hallway to encourage people to ask you questions and submit their ideas from time to time.

8. Free drawings for patents or invention disclosures?

Use royalty-free drawings to speed up your invention disclosures.

A picture or sketch can explain more than a thousand words. The average user may not be keen to draft figures. So, the chances of lost time or ideas are high because people don't want to take the time and effort.

Therefore, it would be handy to have a drawing library, especially for inventors drafting their invention disclosures or preliminary patent applications.

You could build internal resources that provide your product line ups, essential parts, and share them within your organization.

Alternatively, you can use external resources that provide royalty-free drawings.

Resources:

Dimensions.guide (*www.Dimensions.guide*).

Pixabay (*www.pixabay.com*).

9. How to solve the problem, to solve a problem?

Analyse your problem to understand it properly.

A "problem to be solved" is a term used by IP professionals. Sometimes, we are caught up in technical solutions or paradigms of our industries.

Usually, one problem is part of a hierarchy of problems, so it could be that you are solving a greater problem as well.

Key questions:

- Are we solving the correct problem?
- What else do we solve?

In a nutshell, try analysing your problem first to understand it properly. Then, try to make the problem as general as possible and narrow it down to be more specific.

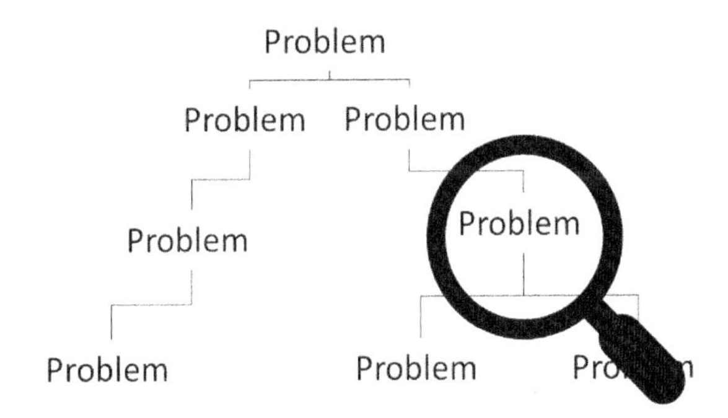

10. The rule of two – for facilitating ideation sessions, inventor interviews, and others.

How to prompt an effective ideation session / IP Workshop.

An IP Manager often has to facilitate ideation workshops, give presentations, train people, or interview inventors.

Ideally, you should always have a second person to support you, even if this person is not from your IP department. This person can deal with technical or administrative issues. This person can manage the operation of the presentation or interview and, thus, keep free the back of the presenter, facilitator, or trainer. That sounds so obvious, but it helps on so many levels.

The overall experience will be better, more ideas will be generated, and more observations will be made. You train a second person in the process, who may become an IP manager in the future as well, or at least you have created an ally within the organization.
WHO are the open-minded people that you can involve in such events?

11. Your decision: Too many or too few inventions.

Your invention disclosure form/portal determines the number of ideas.

Usually, most employees are not aware of the difference between an idea and an invention.

- Idea: Something we could do or should have. A wish or requirement for something desirable
- Invention: Something to achieve what we want to do or has a solution for it. An invention provides an effect that we could use to solve a "wish".

A disclosure form is often used to provide employees with a template to submit their inventions. And some of them submit their ideas, instead.

There are two things that you can do when that happens, and they are the following:

- Use your invention disclosure form to filter out ideas with more specific questions, clear rules, and reject if the invention is only on at the idea stages. This results in less effort and is applicable if many submissions are present.
- Match the idea giver to an engineer, who will help to implement the idea into practice. As a result, you will obtain more inventions. The effort is significantly higher.

You tune your systems to your preferences.

PATENT STRATEGY

PATENT STRATEGY

the fundamentals

12. Why should you apply for a patent?

Understand the actual needs of your clients by asking questions.

I spoke with a fellow founder, and he asked me for my opinion about his rejected software patent application.

I told him I don't provide legal advice. I just asked him what he wanted to do with the patent? Why did he go for a patent?
He looked at me, this bittersweet moment of silence and insight conquered the room.

I said: "If you just need a "patent pending" for marketing purposes, you could ask for a German patent application and never apply for an examination. It will be approximately seven years of patent pending. That you should ask your patent attorney for details."

He was heavily nodding and wanted to call his attorney.
Please keep in mind, that most people don't have a holistic picture of different purposes for patent application. So, we have to poke them with questions, which help an applicant to be clear about their motives and set up corresponding expectations.
Stakeholders need to be taught about the fundamentals of IP. Sometimes, even Managers are not aware of the consequences and existing options besides patents.
Therefore, it's always good to clarify first their motives.

13. How to decide if you should patent an idea or not?

Maintain your patent application quality.

Invention disclosures can sometimes be seen as "chords work." You will receive a lot of them in various quality or relevancy. Some organizations also need to achieve their set goals for filling applications. Therefore, rating and prioritizing such ideas help to ensure that you file only valuable assets in your portfolio. This procedure also offers the chance to meet your annual goals, if you don't have enough high-quality filings.

You can use a rating tool, which is used to score ideas. Those scores indicate whether it is reasonable to obtain an IPR, publish as a defensive publication, or abandon an idea.
It's always recommended to rate inventions in a committee with stakeholders. Perhaps, you can hold a bi-weekly meeting to review new submissions.
The conversation should be inviting and if people disagree with the majority, ask them what their reasons are and vote again.

Example rating fields and their criteria:

- Business alignment (Weighting factor of 4)
    - Core / Future core / Non-core invention
    - Competitor relevancy
- Enforceable (Weighting factor of 2)

    - Detectable
    - Existing workaround solutions
- Patentability (Weighting factor of 1)
    - Assumed Novelty
    - Other applicable rights
- Technology (Weighting factor of 1)
    - Attractiveness of technology / solution
    - Feasibility

The rating scale for each category are as follows:

- 0 = Not applicable or not fulfilled
- From 1 to 4: Less likely to highly probable or clearly fulfilled

All values are summed up and then divided by 6. The final result will then be a number that can be used to prioritize the ideas for filings.

Rating fields	**Rating criteria**	**x**	**Invention A**	**Effective score**
Business alignment	Core / Future core / non-core invention?	4	3	12
	Competitor relevancy?	4	3	12
Enforceable	Detectable?	2	2	4
	Existing	2	2	4

	workaround solutions?			
Patentability	Assumed Novelty?	1	3	3
	Other applicable rights?	1	2	2
Technology	Attractiveness of technology / solution?	1	2	2
	Feasibility?	1	4	4
			Sum	**43**

14. A patent does not protect a business by default.

Two questions to be raised when discussing your patent application draft with your lawyer, which will save you money.

Your patent attorney knows how to draft your inventions for getting a grantable patent. Make sure that they are also protecting your business case. Sometimes, during the long phase of the application, the goal becomes unclear.

If your application is about a core invention or your budget is tight, you can use the following questions to ensure patent-worthiness from a protection perspective.

Thus, ask your patent attorney when discussing a patent application draft with the following questions:

1. How could a third party infringe on my claims?
2. How could I detect such an infringement?

It's imperative to understand if your core patents are enforceable. Otherwise, you create expensive prior arts and disclose your technology to your competition.

You need to understand how to enforce your patent, not just your patent attorney.

"Control your own destiny or someone else will." - Jack Welch

15. Not patent worthy ideas? Do not trash them!

What to do with not-so-popular inventions?

Some technical ideas are interesting, but perhaps not patent worthy. What to do when you can't make that invention free of use for the inventor?

Sometimes you can leverage those ideas and retrieve some value from them.

The following is a short list of possible approaches:

1. If you have the resources, make a national application without request to examine to block /confuse competitors
2. Publish a defensive publication to a public database to create prior art for this case.
3. Print a book via print on demand services and publish it there if you don't want to hint competitors.
4. Publish as a document within your systems as evidence that it was generated, and you may leverage on that idea on another day.
5. If possible, can you disclose as additional description within a different filing?

Public disclosure database:

www.tdcommons.org

16. When to decide for defensive publication?

A checklist to help your decision.

In a typical IP assets generation process, before (outside) IP counsels draft an invention, it is scored (one approach as proposed in Hack 13) and reviewed with an expert community.

Usually, we have two options to evolve new generated IP assets if filing for an IPR is not worth the investment, either make them to be managed (or unmanaged) trade secrets or publish them as a defensive publication.

Therefore, here is a short questionnaire to answer if defensive publication is suitable option for handling your inventions:

1. The subject-matter is not patent-worthy or patentable?
2. It won't create harmful prior art to future filings of ourselves?
3. It does not contain information that could be handled as a trade secret?
4. I don't want this specific topic to be patented by anyone else?

17. How to operate faster than your competitors?

How to decrease turnaround times of generating invention disclosures.

As you may know on your daily IP basis, priority dates matter. Ideas are free to everyone, and those who file their inventions first will obtain legal rights.

From my experience, I see that there is a lot of time spent on the completion of invention disclosures and the preparation of patent drafts.

Here are some of my thoughts on how to shorten the processes:
Provide some basic information and knowledge about the drafting process to the inventors before you start the process.
Use collaboration tools to edit documents simultaneously or use the commenting function with a clear call to action.

Instead of sending documents back and forth, book a two-hour working session with the inventor(s). This is enough to depict the invention and provide an outline for the draft.
If you do not have enough time, request a resource (for example: hire a trained engineer who can do these tasks for you).

18. How can you prevent your competitors' patents from being granted?

Analyse, ideate, and publish ideas and concepts first.

Patentable solutions have to be new and inventive. You should have experts who use methods like TRIZ to anticipate developments and block competitors or market players.

Also, plan and budget at least one dedicated ideation session per year.

So, what should these experts do during the ideation session?

- They put themselves in the shoes of the competition.
- They search for technical problems to be solved, as well as undefined demands.
- They use methods like TESE of TRIZ to forecast the next set of iterations of the competitors' products.

These sessions' output, ideas, and concepts, even if they are only roughly described, can be published immediately as defensive publication, or filed as a patent if you see fit. Please carefully review them to ensure that you don't block your own organization.

With this method, you will create a prior art database that will suit well in oppositions and litigations.

Resource:

https://www.triz-consulting.de/shop/triz-fachliteratur-buecher/trends-of-engineering-system-evolution/

PATENT STRATEGY

patent budget monitoring

19. Find the right patent attorneys.

Invest in patent attorneys who teach you, not vice versa.

If you need to teach the basics about a technical field to your Patent attorneys, find another one quickly!

I had a talk with a befriended Start-up. Their experience with their patent attorney was that the effort is too long, too costly, and their technology is not being understood.

Some patent attorneys are also quite nice, and I think there are very few things on earth that are more complex than patent laws. I believe that the high hourly rate is justified for experts in their fields and specific technologies. Otherwise, the hourly rate is not justified.

If you have to explain everything about the subject matter, this patent attorney is not the correct one for you. Patent attorneys are supposed to be subject matter experts. If you need to explain the basics for the sake of teaching, the patent attorney won't be able to provide you with the best solution.

In such cases, experience really does matter.

Sometimes, your patent attorney asks you common things and basic topics to ensure that you are both on the same page. That would be different.

The issue is to find the right patent attorney, and that takes time. Please start your sourcing as early as possible. Depending on your subject, I recommend having a mix of law firms and solo practitioners. Both have advantages and disadvantages, but it will help you to have the best of both worlds.

What is required:

- Patent attorneys spend time to be introduced by you into your technology and business model.
- Ensure that you have access to a variety of patent attorneys with different backgrounds.

20. How to deal with a clear "it depends" and a strong "perhaps"?

(IP) Legal opinions are never clear as ice, but you can learn.

I often hear from some stakeholders that legal opinion is useless, "can't they just say yes, and no?"; "Is it novel and inventive?"

That's the issue with very complex environments in the IP space. Often it depends on circumstances that we cannot anticipate or which a single person cannot fully understand.

Each examiner, judge, or attorney can interpret certain decisions differently. Each jurisdiction has its own rules and specialty. Hence, it is hard to make any predictions and it consumes lots of financial resources.

I think that it's only after the very end of an opposition process of a patent that you will understand if a claimed invention is really considered novel and inventive.

My advice to you is: if you receive an "it depends" answer - just ask the 5 Whys: Five times ask the questions "Why is that so", and/or "Why it depends". If you cannot depict it from the letter, ask the attorney.

Attorneys may dislike that. So, be aware an attorney will probably charge you for their answers. With that technique, you get a pretty good understanding of why it's seen like this.

There are very few absolute statements in IP. Try to understand more the background of any statement and be aware that any statement can be wrong, even if they come from an office.

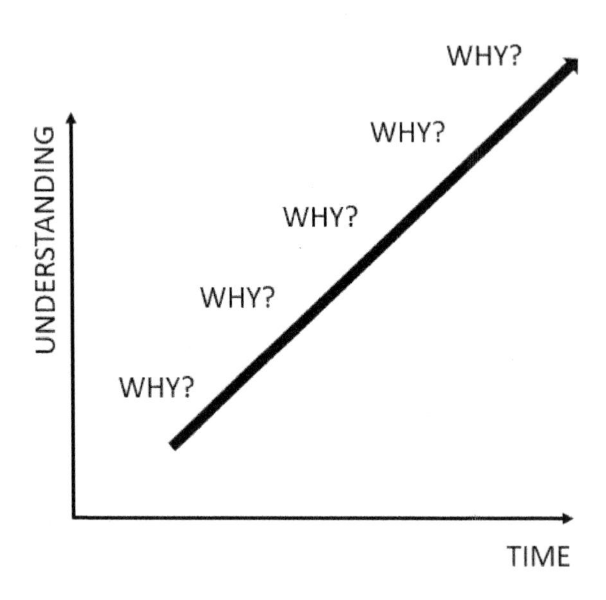

21. How to reduce costs in your patent application with minimum effort?

Cut costs with your patent applications by adding blank drawings, which can be used for patent applications.

I spoke with an outside counsel of one of my clients and asked them why they don't use the beautiful, crafted patent drawings of my invention disclosures. They were in black and white, done with the greatest care, and looked amazing, in my opinion.

He answered, the drawings are fine, but it's the reference signs, and lead lines are the point. We need to arrange them differently. Sometimes it takes too much effort to erase them, and that's why it's cheaper to redraw the drawings with the help of a draftsman.

As an immediate and simple action, I added an "Appendix" to invention disclosure documents, which contained blank drawings without any reference lines or reference signs and made it as part of the delivery.

Since then, the drawings have been used. It saved my client a lot of time and money.

My advice is the following:

- Add an appendix to your invention disclosure form when reviewing an invention disclosure.
- Ask the inventor if they could provide the drawings also in black and white.

22. Hidden costs in your patent portfolio – Patent renewal fees.

Cut out the middlemen.

You must pay annuities to renew a patent in each jurisdiction. Soon, this will become very difficult. That's the reason why most companies either use a law firm or a service provider to pay the renewal fees.

What people don't know is that there are often unnecessary service fees hidden in their invoices. This can cost you several percentages per year, and from the experience of service providers, this could be up to 40% of the total invoiced amount.

Back then, in the old days, only law firms could perform such renewal fees payment. Some law firms could make their living just from this service alone.

Back then, in the old days, only law firms could perform such renewables. Today, anyone can do it.

Therefore, it is recommended to carefully review the invoices and compare them with other service providers. There is no need to sponsor a law firm. Most service providers offer cheaper delivery, and in fact, these providers are used by the law firms as payment providers themselves.

The following are some providers:

www.patentrenewals.com

www.pavis.com

23. Is it a must to pay for IP activities from your own budget?

Do not pay for the cost of your IP activities by yourself. Use governmental subsidies.

The budget is always going to be tight.

Usually, you have more inventions or ideas than the budget allows. One strategy is to use your budget more efficiently, and another strategy is to prioritize the inventions.

There is also a third strategy, which is often overrated: Have you leveraged all applicable subsidies and other financial support programs?

For example, if you fall into the German jurisdiction, you could cut half of the cost of your patent. Please reach out to the responsible decision-maker and make deals. Get some fundings and use the free cash to support your IP strategy.

The applications can consume some time, so you might want to think about hiring people or working students to do this for you.

Please note, there are specialized service providers that can help you with subsidies.

24. Rejected by a patent office?

How to react, generally, to non-final rejections from a patent office.

You can't prevent letters or opinions of the patent office.

However, we can manage our reactions to it. Following are some alternative perspectives and my advice towards such rejections:

1. Be ***happy***. It seems that you have at least tried to receive the broadest scope of protection.
2. Be ***kind***. Examiners are not your enemies. They are interested in granting your patent and want to help you. So, work with them, not against them.
3. Be ***clear***. If the examiner interpreted your application wrongly, maybe some others would do the same. Look at this as a chance for learning.
4. Be ***open***. If you haven't already got a professional advisor, get professional help from a patent attorney, or get an additional subject matter expert involved.

Also, start to collect data which could relate to the reasons and circumstances of the rejection. You may spot certain dependencies or causations that help you to improve your fillings and reply to office's decisions.

PATENT STRATEGY

patent analysis

25. The differences among patent families, extended patent families and publications.

Developing an IP reporting policy.

Patent gives an exclusive right to prevent others from exploiting your inventions without permission in the granted territories. The publication of a patent application will be carried out within eighteen months after the filing.

A patent publication is a document which was submitted to a patent office. One invention creates several publications over the prosecution period. Usually at least one patent application and one granted patent.

Patent families are groups of patent publications in multiple jurisdictions, which refer to the ***same*** invention. And extended patent families are groups of publications, which refer to the ***similar*** invention.

The big problem is who decides which publication belongs to the same patent family. Patent offices, database providers, or even the organization have different calculation methods to count them.

Why is this important to consider?

Because when you have to report, the numbers will be different depending on your data source or your selected form of counting.

Therefore, provide an IP reporting policy where you explain how to count and which data to use.

Furthermore, for your own patent families, collect the family IDs from different data sources.

26. How to perform an invention (product) to IP-rights mapping?

Leave a trail or record cleaner than as you have found them.

The two following questions will always be asked to you as an IP Manager:

1. "How many patents do we have in this field ...?" or
2. "Which patents are related to this product ...?"

And, you have to answer. But ***how?***

You need to map inventions or products to corresponding IP assets or IP rights.

I have seen three approaches to deal with that question in my time with EVOip GmbH and they are the following:

1. I don't have any tracking at all. It will be solved with the second approach when necessary.
2. From time-to-time, inventions- and products-mapping to IPRs are desired in a project or upon request. Usually, this process takes up a lot of effort, dealing with bad or outdated data quality, and overtime hours.
3. Continuously tracking / updating when generating or maintaining IP assets. This is the most convenient way, with low to medium effort.

My advice is hence, start with the third approach and add this to your standard routine to either spend one hour each week updating your records, or revise your records every time you touch them.

If you do not have the time for that, at least add a marker or tag for the records that seem odd or need revision.

Thus, you can understand which records in your system you can trust or not.

With that you are leaving your IP trail much cleaner and easier to follow over time. If you are then asked any questions about your portfolio, you can run along your IP trail without the fear of getting lost.

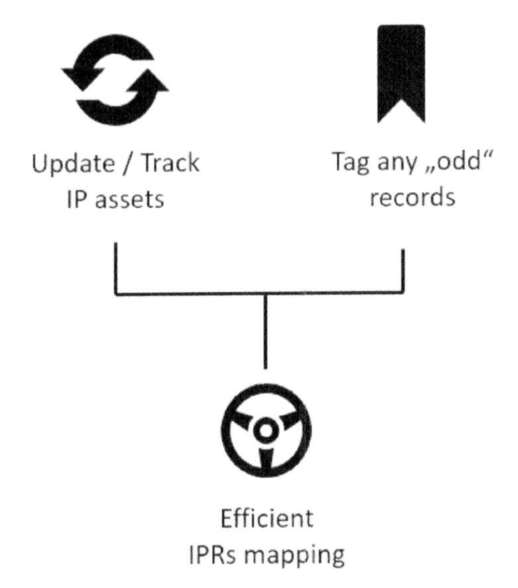

27. It ruins your patent analysis!

The secret phase in patent prosecution will deliver a false perception.

In this hack, I want to share the Fog of the secret phase and how it ruins your patent analysis.

Sometimes, we must provide simple trending analysis based on patent data – that were taken most recently and ASAP.

Recently, I saw an S-curve for a certain technology in a field that started flying, quadrupling the filings within the last two years. "Suddenly" the chart of the second half looked quite different -- like a decrease.

There is one aspect which some stakeholders are not aware of: *The secret phase*. The secret phase of a patent document goes on until the document is published. This phase goes back around 18 months.

So, most charts will show a decrease in applications for the last 18 months. In the next year, the very same chart will look different because all patent documents will be published, and the data will become available.

How could we solve it?

- Visualize with a trend line based on the last three years. Highlight this line with a different color to demonstrate that the last 18 month-period is questionable.

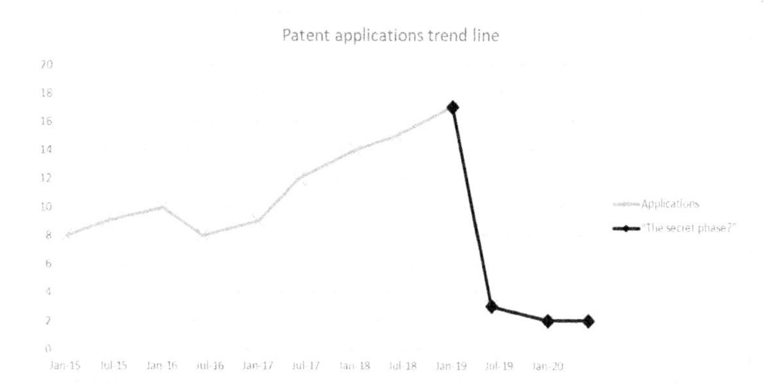

- Alternatively, remove the current year and last year's data.

Otherwise, your chart could be misleading, and follow-ups could be misinterpreted.

28. How to rock with your patent reports.

Provide a summary as the first page and a clear call to action.

Only one person would enjoy reading a patent analysis entirely - the analyst who wrote it.

For everyone else's sake, please provide a summary as the first page.

In my mind, IP Managers are busy and have very little time. They prefer to have the information in an easily accessible form. Any professional document is not a novel, where you need an attraction keeping arch over the document.

When you review your documents, check if they are set up for the following requirements:

- Easily accessible key information?
- Clear statements?
- Handy format to process your work for the next step in the process?
- Based on solid data?

My advice: Ask your circle of trusted peers to exchange their best practices and exemplary templates and take the best out of them.

29. The magic of patent portfolio analysis and how to make it pay for itself.

Analysis for pruning your IP portfolio and save costs on renewable fees.

As always, IP budgeting is tight. And an external patent categorization/ portfolio analysis can be costly.

My recommendation is always to use your patent portfolio analysis in order to prune your IP portfolio and cut costs on renewable fees. So, it is clever to start a list with patent families you should review.

Basically, each alive patent family with an age of 15+ years is worth be reviewed by default:

- Do you still need to consider all countries of the protection?
- Is the technology still used, and is there still no other alternative available?

It only makes sense to put such families in discussion along with the input of your R&D and Sales department.
R&D really knows best about the value of that technology. Sales know the best about the competition or market impact.

For large IP portfolios, you could save, at the very least, thousands of euros with this simple approach. A single patent family may cost you several thousand euros in the end for each protected year. The savings could be used for new applications or financing ideation to create more IP assets.

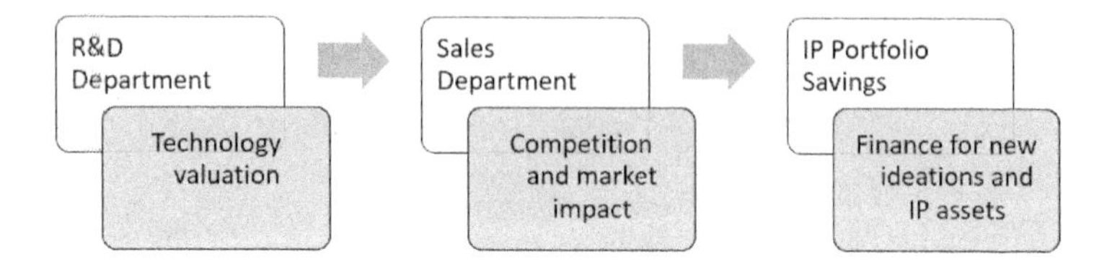

PATENT STRATEGY

patent search

30. How to make false hits in a patent search your friend?

Do not avoid the "NOT" operator.

In general, patent searchers try to avoid the "NOT" operator due to the possibility that this operator may exclude relevant results.

Still, the "NOT" operator is your friend.

In your search, you will have some false hits, which will falsify your quantitative analysis. No matter how detailed your search strategy becomes, usually, some hits are not useful. The most obvious step is to delete them manually or redraft your search strategy again and risk excluding relevant hits.

This, however, does not solve the issue in the long run.

My recommendation is the following process:

1. Build a list of false hits or patents that you consider after reviewing their relevancy.
2. Generate a "NOT" list.
3. Identify irrelevant patents within your search result and potential similarities they have. Also, add the patent families to your "NOT" list.
4. Subtract the original search strategy with your "NOT" list.

You can create "NOT" lists for your clients, which help them exclude patent families and avoid repeated work in the future.

Alternatively, you can update your search strategy with "*proximity operators*". Instead of searching through the whole document for the presence of keywords, the proximity operator limits results to a smaller distance, so that your keywords are found in correlation to each other.

31. Function oriented search is the patent searchers friend.

The concept of abstraction in the TRIZ functional analysis.

Sometimes patentability or invalidity searching is a bit like standing in front of a wall. You don't find any references. What should you do?

Jump over it.

One aspect is the concept of abstraction in the TRIZ functional analysis, and that's what you could do in this situation.

Function Oriented Search of the TRIZ is equivalent for technical concept searching.

Basically, you don't search in patents for the physical characteristics that the invention has, but more about the objective tasks, the functions, or technical effects provided to find workarounds or solutions to your technical issues.

Derive from the specific functions like "feeding gasoline to the fuel injector" to a more general function like "moving liquid."

There you may find references in other technical fields that you could cite.

Please see this document from Tiziano Montecchi and Davide Russo:

https://www.sciencedirect.com/science/article/pii/S1877705815042472

32. How to FTO in FIVE steps.

IP Management is more than only a legal challenge. By Dr. Claas Gerding-Reimers.

Technology companies *that* intend to market innovative products, face the following situation: 3 million patent applications are being filed, yearly.

So, how to practically deal with the legal risk of potentially infringing millions of patents as a business executive. How to assess FTO?

"Let's ask our external patent attorney, if *we have FTO for our innovative product", one might think. Sorry, that does not work! An attorney is qualified to give specific legal advice but not business management advice. FTOs are more than a legal challenge,* they involve *resources, risk aversion, strategy, agility, economic consideration, etc.* matters *to your business, as well.*

Generally, FTO processes should be designed to not inhibit agility in innovative product generation. On the other hand, legal business risks have to be managed, as well. In relation to "innovation", "FTO" would often be in a sensible trade-off position: **Innovation agility vs. legal risk management.**

A possible 5-step process to smartly tackle "FTO" in a nutshell looks as follows:

1. Check your business case for your product (chances)
2. Find out which legal/economic risks you are willing to take with your product (worst case scenario, etc.)
3. Decide on your FTO resources (manpower, money, time) to spend. Base your decision on step 1 and step 2.
4. Do a reasonable (in sync with step 3-resources) IPR/patent search and evaluation. Stay agile!
5. If necessary, ask an (patent) attorney for legal advice on potentially relevant patents identified in step 4.

One is well-advised to design FTO processes and perform FTO tasks in alignment with the business and in conformity to the IP strategy. An experienced and business-focussed IP manager understands the multifaceted challenges of an FTO and helps you to succeed.

Read the full blog post version on www.1stclaasip.de:

https://www.1stclaasip.de/fto

33. How to spot a good Freedom to operate search?

Four criteria to spot a good FTO search.

At some point, you will receive patent searches for Freedom-to-operate purposes. So, how to spot if the Freedom-to-operate search is a good one?

A good patent search for a Freedom-to-operate should match the following criteria:

- Be ***systematic*** through features analysis and match the features with corresponding patent classes.
- Be ***comprehensible*** through a disclosed search strategy and search history.
- Be ***consistent*** and have clear criteria for what is relevant in case of doubt, and that the right remains in the report.
- Be ***helpful*** by identifying the most probable and relevant IP rights and mark them.

A good test is to double book a Freedom-to-operate search and compare results. Start writing your own specification document in which you address expectations from an FTO, then hand it over to your suppliers and in-house patent searchers.

34. How do you know when you are done?

You need to stop at some point.

Kai H. Simons, the Agile & Scrum Coach, inspired me today to talk about the ***"Definition of done."***

I have the feeling that in our IP field, there are a lot of unsaid expectations for work results. A result of a "patent search" is very diffuse and could mean a lot of different things. When is an application done for a patent?

Especially in a prior art search, you could search for eternity. You need to stop at some point, to know when it is enough. Often, this is triggered by the experience of the searcher, and different people will then stop with different results or after taking some effort.
A *definition of done* is an agreement within the team or between client and service provider on whether a certain task is sufficiently fulfilled with completed/delivered actions.
A *definition of done* boosts quality of work and prevents insufficiencies.

"If I want to get work done, that's usually about 3 in the morning."
- Marc Andreessen

IP and IT

35. Why do you need to know about "Shadow systems"?

Hunting for shadow systems to boost productivity.

Shadow systems are secondary places where we store information. For example, Outlook (personal mailbox), file drives (as a file saved), a data table, a notebook, etc.

A lot of users build their own shadow systems due to the lack of availability of better systems, knowledge, or for the sake of convenience.
Sometimes, this becomes our preferred working style. However, in some cases, it happens that we have to track more information than what our software allows us to do.

Also, this could make it challenging when we take over the duties of other people (who have developed the shadow systems), as their systems become irreplaceable and they are perceived as "experts".

I admit, I personally have three (shadow) systems. An IP-management tool, a shared Excel table, and my mailbox.

I can't save everything in the form of files or information in the IP-management tool. That is why I added two shadow systems for my reporting purposes.

If you need to take over the duties from a person, remember to ask for any shadow systems that they are operating.

If you are leading a team, identify these systems and do a root cause analysis. Often, some shadow systems are no longer required or are duplicated by multiple users.

Therefore, that's when it would be a great opportunity to increase productivity and lower the dependencies on certain people.

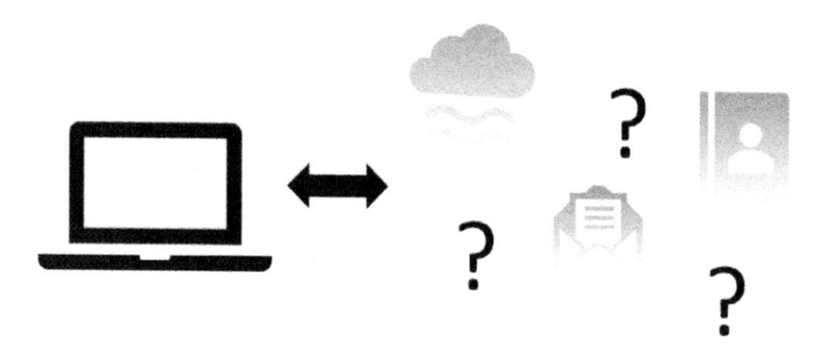

36. Do we need IP Management software?

You don't need software; you need a solution.

"Do corporations really need IP Management software?" – It was a question from Dr. Robert Fichter.

Perhaps I (and Robert) will become the enemy number one of IP Software companies by asking this, but we should keep the narrative in this direction.

In my opinion, I could imagine that it may be attractive for anybody if they could avoid learning new IP management software.

However, I just believe that this can be very dangerous.

Let me be clear: You need to manage the Intellectual Property aspects of the business, depending on your context.

It's your responsibility to ensure that work is properly done, documented, and requirements are met according to suitable standards and best practices.

You can do this without software, but it will just be more challenging. Please, leverage on modern technology. It makes sense to request an official software project, including integration for your organization.

If your management doesn't want to provide you with suitable software support, calculate the saved cost through increased productivity and compare the total cost to a single lawsuit that you could shorten or prevent with the help of software.

37. The real pain point of IP Managers to be solved by software.

Software tools will make your life easier -- not complicate it even more.

One day I had a nice talk about IP management software and the overall branch's experience.

One question raised was: What do you think are the most impactful pain points for an IP Manager that good software should solve?

I think the criteria to be included are Time, Budget, and "Convenience."

Time: I have never met an IP Manager who had enough time to do their job. So many of them put so many extra hours compared to regular employees.

Budget: They never have enough budget for all their cases, so they need to decide and take responsibility for their own decisions.

Convenience: Often, they have no or only a few assistants to do manual copy-paste work of inputting data and have multiple systems to manage. Those systems are a nightmare in terms of user experience.

Good software should tackle these points. It's not enough to provide a system to manage work. A system must provide value.

I sincerely appreciate highly integrated and adaptable tools which can save time and make our work easier.

Here is a selection of software categories you may consider:

- Professional IP Management tools
- Professional Patent search software
- Professional Document management software
- Professional and up-to-date Office software

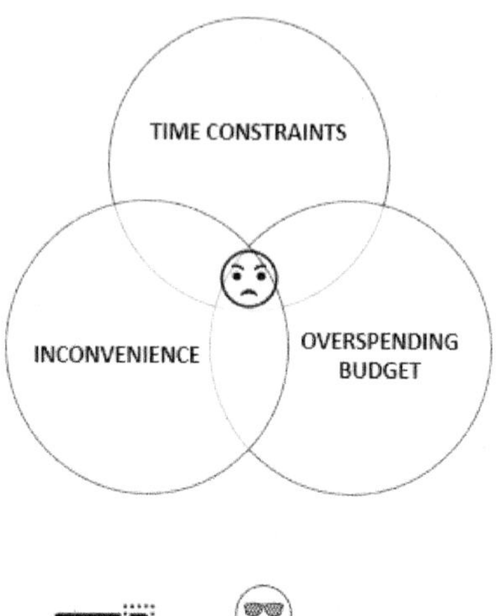

IP MANAGEMENT

IP MANAGEMENT

the general

38. Is your IP Management a black box?

Use customer service's best practices.

Some people in the organization may perceive an IP Management department as a black box. A black box is something that receives an input, and somehow, creates an output. Sometimes with the same input even a completely different output.

Such a perception is never good. At least, if it is not true.

You need to do "IP-Management marketing" then as follows:

- The Head of the IP department must be the spokesperson for IP Management.
- If you have defined processes, make them visible to stakeholders.
- You need to reason with your decisions and inform stakeholders about those decisions.
- If you rate inventions, inform them about the criteria.
- Use the opportunity to inform managers about your work, actions, and inputs you did.

You are a professional, and professionals don't work by accident - they have standards, processes, and procedures.

39. IP or Patent Management?

Not everything is about patents.

Sometimes we may become too focused in our IP Management on a specific field. And usually, we are focused on patents.

Be aware, there are other IP fields that must be taken care of as well.

The following are some examples:

- The likelihood of a copyright infringement is at least ten times higher than a patent infringement.
- Sometimes you could also receive design protection.
- Trade secrets need to be managed and protected.
- Are the trademarks sufficient?
- The collected data needs to be treated and protected as intangible assets.

Review your IP Management activities and see if you detect a cluster. If you take into account the context of your organization, does your IP focus adequately on business goals?

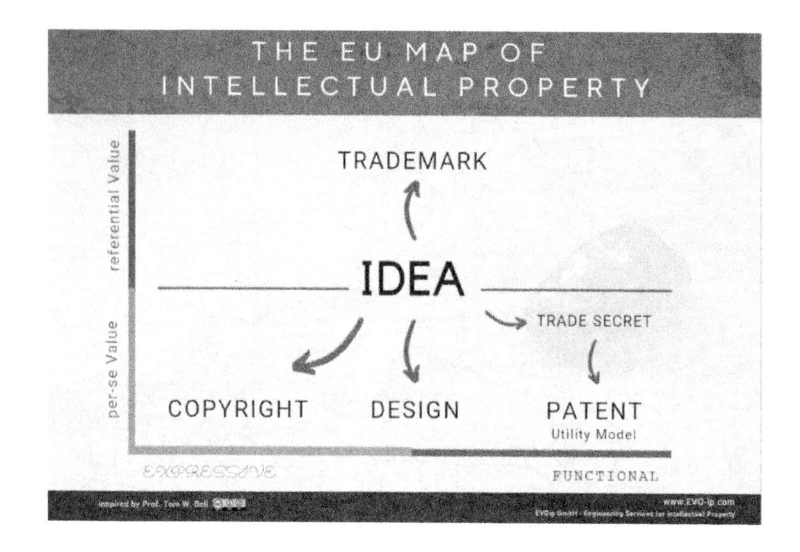
THE EU MAP OF
INTELLECTUAL PROPERTY
referential Value
per-se Value
TRADEMARK
IDEA
TRADE SECRET
COPYRIGHT
DESIGN
PATENT
Utility Model
EXPRESSIVE
FUNCTIONAL
www.EVO-ip.com

40. Does agility apply for IP Management departments?

Agile methods to prompt effective IP Management.

What do the IP Management department and Agile Project management have in common?
Could there be some similarities between the two groups of IP and IT Professionals?

In fact, there are some similarities:
People of these professions are usually pretty senior, clever, and well-educated. Both groups deal with the most complex intangible subject matters, which are often hard to estimate in terms of effort. Both, sometimes, have to react to unplanned events. Both have customers with (hidden) requirements. And, both have to deal with deadlines and need to provide values.
As an IP Manager, I like to think that we can be inspired by software developers, especially with regards **to self-managed teams, common planning, and visual management (Team boards)**.
Last but not least, both groups find pretty clever ways to deal with similar challenges. Below is a good read with a lot of applicable tools from Management 3.0.

Resource:
https://management30.com/

41. Intellectual Property Manager or Internal Politics Manager?

Excellency in IP is silent.

My clients are mostly large corporations, and I am, every single time, amazed how much they have to deal with complexity. Not technical, but political matters.

I would like to highlight that. Because sometimes in the IP department, you just want to get the job done.

Instead, you have to take care of forbidden words, charm with other departments to receive engagement and meet your KPIs. For your own self-advertising requirements, you need to do things that might not add value to the company.

So, the silent work of a good IP Manager, such as carefully managing expectations and adding value to the company, is often not visible.

This hack is to apply project management techniques like stakeholder identification and -management in your work. Define your own policy to interact with colleagues, superiors, and other stakeholders. It helps maintain your own standards, which should be corresponding to the corporate culture.

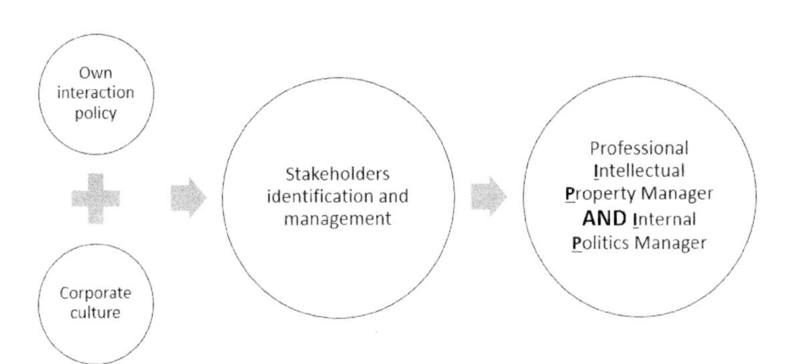
Own interaction policy
Corporate culture
Stakeholders identification and management
Professional Intellectual Property Manager AND Internal Politics Manager

42. A sign for bad IP Management?

Make the effort to understand the business.

An obvious sign of bad IP management is the absence of any understanding of the business of an organization within the IP department. You can check that by reviewing the amount of time spent on understanding the business – per year.

Usually, there is none or little spending on visiting production plants, fairs, or other introductory activities for IP Managers.

IP Managers must know the business from end to end. The business is always evolving, and IP Management must catch up with it. Otherwise, bad decisions are made, or opportunities are missed.

Keep in mind these 4 How-to's:

- Schedule annual factory visits.
- Visit internal product demonstrations.
- Cross all departments and work with them for one day as IP Managers during their onboarding procedure.
- "Create" and assign IP ambassadors in business departments and organize a meet up.

43. Key ingredients for successful IP Management.

Four factors to achieve success in IP Management departments.

Usually, achieving success in IP Management comes from having a great team, skills, discipline, and knowledge.

All these domains are crucial from my observation of great IP departments.

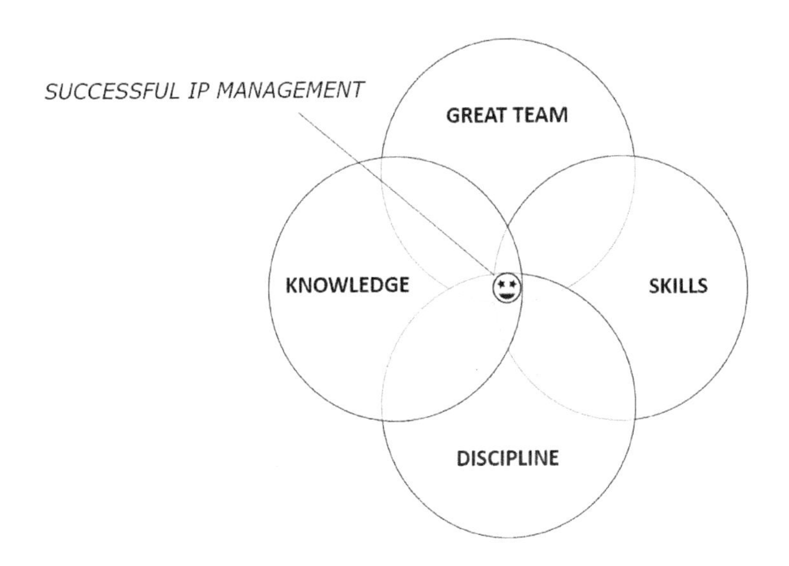

Great team: Actually, people like to work together, and they generally have a positive attitude towards work and colleagues.

➔ *Therefore, hire first for attitude over skills.*

Skills: People are capable or trained to perform necessary tasks. Constant effort is always taken to sharpen and broaden skills of the people in the team.

➔ *Therefore, develop a mandatory training plan, for your employees, and for yourself.*

Discipline: Discipline is necessary to face the challenges, due to the nature of the intangibility of the subject matter. Discipline is what is left after motivation is gone. If you need to rely on people, ensure they are disciplined, and you are demanding the appropriate level of discipline. People will act in the way you empower them to act.

➔ *Therefore, discipline is something you need to require.*

Knowledge: Knowledge is imperative. Based on your data, records, notes, and available information, you have a strong influence on the productivity, quality, and effectiveness of your IP Management. You need to find ways to maintain, increase, and distribute the knowledge within your department.

➔ *Therefore, look for professional IP knowledge management*

IP MANAGEMENT

IP awareness and IP knowledge

44. Raising IP awareness?

Why it can be good to be sued for a patent infringement.

I had an inspiring exchange about IP strategy and DIN 77006, which crossed on the topics of IP awareness and current cases of potential patent infringement in my mind. Such legal occurrence immediately increases the IP awareness of the affected company.

That made me think of an old military mnemonic:
"*Affectedness creates awareness.* (Betroffenheit schafft Bewusstsein)"

Please do not risk patent infringement. But in case you are affected, you might use this to create an IP awareness that will be long-lasting.
Other recommendations for creating IP awareness are the following:

- A personal encounter with an IP Manager as a part of the onboarding process of managers.
- Scheduling regular training, which is carried out by someone "inspiring" or a knowledgeable special IP guest.
- Special events such as Inventors' days, Inventor awards, study trips to the EPO.
- Creating supportive networking of the patent department within the company.

- Training of IP ambassadors in every major department of the organization.

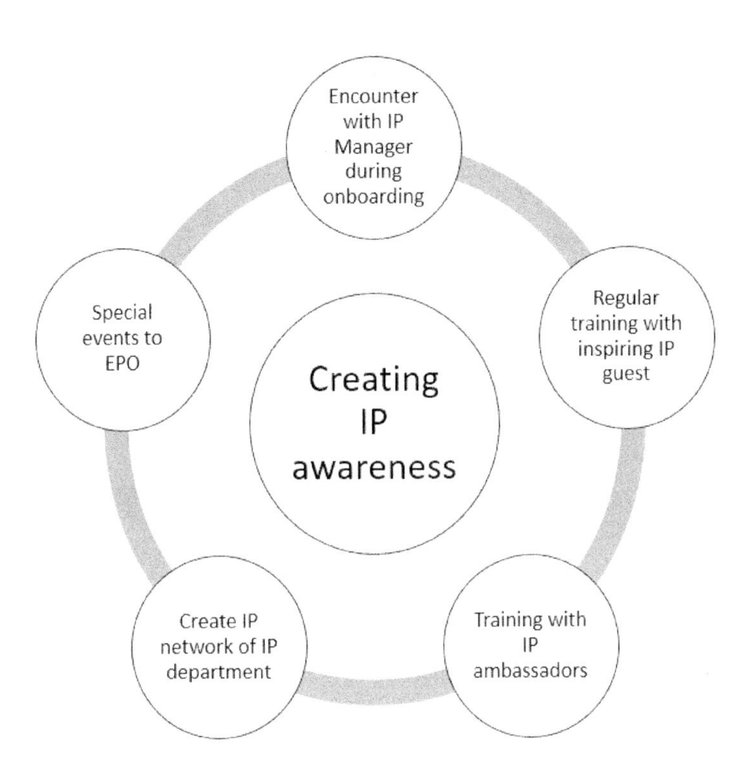

45. Knowledge management of IP assets

Do you protect IP know-how in your IP department?

I think one of the most overlooked areas in IP department is IP knowledge management. As IP professionals, we know how to protect know-how. However, it often happens many times, that we work twice and generate same information or data a year later.

So, how could we overcome this situation and emerge new knowledge assets?

We need to provide the right tools and change our behaviour. Following approaches can easily be implemented:

- A wiki page as a content collaboration. This simple function is (till the date of this book's publication) available withing Microsoft Teams.
- A self-explaining file archive, where folders and files' names give clear hints on included content.
- Saving relevant emails to this file archive, so that they are searchable for the whole involved department.

If you haven't thought about this yet, make it a topic in your next department meeting and raise the question: "How do we make today's information accessible and easy to find years later?"

Resource:

https://hbr.org/2015/01/managing-your-mission-critical-knowledge

46. Let's build a quote factory!

A way to foster an understanding of the importance of feedbacks and IP workflow.

Workflow and understanding requirements are a must in IP Management. We are spoiled by the clear rules of the EPC, so sometimes we are not aware of the hidden requirements. I have learned an exercise from Jurgen Appelo which is aimed to raise one's awareness of project prerequisite: ***The quote factory***.

The task of this exercise is to produce inspirational quotes that match the "*client's*" requirements.

The teams have multiple rounds to learn and adapt. 10 minutes to make 50 quotes and 3 minutes of set up time prior to that.

They receive vaguely formulated requirements from the "*client,*" and the "*hidden*" exercise is that they should ask for feedback early. If not, it will be followed by more work. (Welcome to the real world!).

I (as client role) have 8 requirements, which I disclose only two within each feedback request. I tell them my requirements only if they ask why I reject their "products" such as the following:

1. All quotes need to have a name of the quote authors.
2. All quotes need to be written in a sticky note which has a border line in layout.

3. All quotes shall be written in capital letters, except the author's name.
4. Only one quote per sticky note.
5. The quote and the originator shall have different font colours.
6. The author name field shall have hyphens as pre- and suffix.
7. All quotes shall cover a certain topic, e.g., work ethic.
8. All quotes need to be surrounded by quotation marks.

This exercise provides recreation, and people will have some first-hand insights to structure their work and ask for feedback early in the process.

> *"A satisfied customer is the best business STRATEGY."*
>
> \- Michael LeBoeuf

47. IP education in companies?

Methods and concepts for IP education.

What are the common shortcomings of IP education programs in companies?

The following is a shortlist:

- Too boring.
- Too detailed.
- Too time-consuming.
- Kind of schooling teaching method instead of coaching / training for adults and professionals.

We should develop appropriate methods and concepts for IP education like the following:

- Gamification,
- Interactive workshops,
- Direct Involvement of the participants,
- Recurrent Bite-sized information,
- Quizzes and crosswords puzzles,
- and others.

In terms of generating curriculum, for example, create a table comprising the different roles or management levels and define their corresponding levels of IP awareness. Then assign specific pieces of training or points for each role and reach out to facilitators to make your proposals.

48. IP Management Tyranny - UX problem?

Revise your customer experience with your IP Management.

I asked for bad examples of IP Management. And the IP Management Tyranny was shared: I know stories of an IP Management department behaving cocky and restrictive to inventors and stakeholders.

The reported experience of my contact is:
"As soon as your IP Management department gets involved, things become unnecessarily difficult. They want to take the lead of my projects and left me as a bystander of my own project". “The IP department is a necessary evil. We have to endure because it's mandatory.”

If your (or inventors') first impression of IP Management is negative, your IP Management has a serious problem. Did you check you how to do better with your IP Management?

Some questions for reviewing the IP experience are:

- Is a bad experience due to misunderstanding?
- Is the existing IP policy too strict to provide a good experience?
- Are the processes too comprehensive, complex, or cumbersome?

An excellent tool is to see IPM Services from a user experience perspective. I suggest asking the following questions while developing such a tool:

- How does an inventor perceive its interaction with the IPM tool? How does a project leader perceive the replies?
- Is the Top Management provided with the correct amount and level of information they need?
- Is an Inventor / Project leader or Management emulated with experience do the best together with your team?

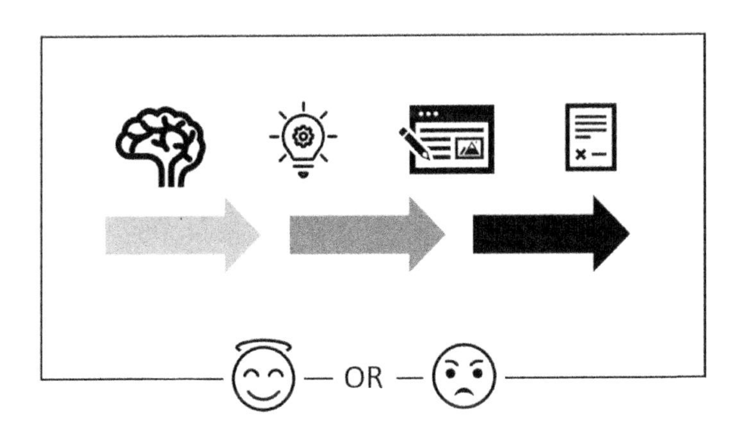

IP MANAGEMENT

quality management

49. Do we need the DIN 77006 Quality management in IP?

Quality management in IP is a must-know.

During a company lunch meeting, we had a conversation about the standard *DIN 77006*.

One statement that was made by a patent counsel of a huge multinational company was: "We carefully reviewed it, but we couldn't consider a value for us."

In fact, he had a point. I am certified as an auditor for Quality Management so please allow myself to formulate an opinion:
The *DIN 77006* isn't a handbook or a guide. If you have professional and effective IP management, it won't serve you as an inspiration or as a must-have.

The reason for that: *DIN 77006* is based on the best practices and is written in an abstract manner to suit all sizes of businesses.

If you are already excellent in IP, there is less potential to learn from DIN 77006. When you have to set up new IP management or deal with ineffective IP management, *DIN 77006* can be more helpful.

In my opinion, the norm will rather help make your IP management processes more comparable among different parties. This allows you to outsource services, develop branch software, exchange expectations, and onboard new IP Managers much easier. Also, it serves as a benchmark for managing IP Assets and ensures proper IP Management.

DIN 77006 is a standard about quality. Hence, it's less problematic to find good quality management (QM) consultants than IP management consultants. So, think of hiring or asking your own QM-Team to support you in systemizing your IP Department. They don't need to become IP experts, and IP experts don't need to become quality management professionals.

Resource:

https://ipbusinessacademy.org/

50. Do you provide value to your clients?

How to make sure your services are professional.

It is usually quite common for law firms to bill hours. In my opinion, that billable hour system is no longer applicable in our times, because there is no incentive for the law firm to work faster.

Also, if every minute counts, clients could spare the necessary conversation in order to save budget.

Would you like to pay someone to listen to your feedbacks to improve their work?

That is not how to provide the most value to the client – who is you. Therefore, it makes sense to agree on a cap or flat price for certain activities or communication.
For instance, you could set up a free bi-weekly governance call to discuss ongoing projects, administrative topics, or feedback.

Professional services are services that create value and have understandable pricing for clients. Before and during the collaboration with law firms, I will take these questions into consideration:

- Do your contracts/partnerships incentivize continuous improvement and communication?

- Does your law firm have in mind how to create value for you or them?
- And in case they create value only for selfish reasons, who can you replace them with?

51. How to maintain service quality?

Four approaches to ensure certain service performance and quality.

This is an insight of my visit to the European Patent Office (EPO) that you could use for your IP Management.

The EPO uses several approaches to ensure specific performance and quality. Quality managers would love organizational set-ups for auditing such as the following:

Firstly, a committee makes a final decision of granting a patent by three people. Two examiners per case and one chairman are appointed to ensure an objective view.

Furthermore, defined processes and communication are required to guarantee consistent results with a unified communication experience for all applicants or stakeholders. Mandatory documentation of a process based on performance must be carried out and stored so that a process can be reviewed, if it is challenged in later situations.

And finally, assign at least one designated, experienced person in each team to assure the desired quality.

1. A committee of 2-3 people for an objective view
2. Define and agree on processes and communication standards
3. Define the scope of mandatory documentation
4. One designated experienced person for quality check-up – EVERY TIME

"Quality is remembered long after the price is forgotten" – Aldo Gucci

IP MANAGEMENT

IPeople

52. Why experience doesn't matter. Better name the masters!

Don't look at experience, look at mastery.

Is that a bold claim? I think it is important to distinguish between experience and mastery. Experience is, for me, just an accumulation of spent time.

Like Kurt Tucholsky said *"Experience means nothing. You can do something badly for 35 years."*

So, I consider experience as a quantitative thing which may be useful or not.

Mastery is a different thing. I see this, especially in the IP field a lot. The people who have deeply developed their skills, knowledge, and competence. They do not only gather experience, but they also leverage on it. They apply it.
So, I don't look at experience, I look at mastery to learn and be inspired.

See examples of great IP Influencers in this post:
https://www.linkedin.com/posts/ricardocali_patents-ipmanagement-activity-6806825112659095552-2jpE

53. What makes a great IP Manager?

Essential qualities of a great IP manager.

Some essential qualities you need in order to be a good IP manager are as follows:

- Disciplined,
- People-person,
- Decisive,
- Knowledgeable in the business,
- Autodidactic,
- Informative,
- Knowledgeable in IP.

I think, as an IP manager you have to be a mediator of people, master of data, self-manager, and information facilitator.

If you need to hire an IP Manager, look for the right attitudes, because it is easier to teach someone the necessary skills or knowledge than trying to change one's attitudes.

The following is an excellent read about the crucial qualities of a great IP manager:

https://www.ipwatchdog.com/2020/02/15/identifying-crucial-qualities-great-ip-managers/id=118856/

54. The seven tasks of any IP Managers.

All IP Managers have strategic and operational tasks to focus on.

This list gives you an overview of your tasks as an IP Manager working in and on an IP Management system.

Strategic tasks:

- Strategic orientation of IP management with business and innovation strategies and deriving an suitable IP strategy.
- Stakeholder management and ensuring IP awareness.
- Continuous improvement of the IP management system.

Operational tasks:

- Planning, execution, controlling and improving of operational tasks (PDCA-Circle - while working on your IP Management)
- Advanced planning and budgeting, especially resource / budget management, and procurement of IP services to be provided or received.
- Conduct trainings to improve or maintain IP and non-IP skills (for yourself, your team, and organizational levels)
- Reduce and prioritize tasks, data and duties (clean out and keep lean)

55. The four core skills of a patent professional?

How to use a skill matrix.

The basic skill set for any patent professional is the following:

- Analyse,
- Search,
- Describe,
- And discuss.

These skills, as mentioned above, are the practices that a patent professional has to be capable of performing. If you need to assess the skills of a candidate, measure and evaluate them in these four categories.

You can define the categories even more deeply, for example: "Analyse -> Analysis of the protective scope of a patent".

Then let the person rate their own capability by using the following scale:

- 0 = Can't do it at all
- 1 = Can do it under direct supervision
- 2 = Can do it by myself
- 3 = Can teach others how to do it

With this simple structure, you can quickly obtain the current skill set of your team and plan pieces of training accordingly.

56. Not another meeting.

Block your (and your team's) focus times.

A classical IP department has many interactions with different stakeholders: Communication with inventors, outside patent counsels, internal meetings, and other appointments.

There is one fact. In a meeting either you work, or you meet.
How do most IP departments still manage to complete their work? I assume by working overtime hours or not paying attention during a meeting. Why then would people send you E-Mails about inventorship during that night?

"You have time, you make the time."- Arnold Schwarzenegger.
You and your team have work to do, work that you need to be focused and concentrated on.

You need to review your schedule. As a manager, you need to have at least ten hours per week of uninterrupted focus time. As a professional without having direct reports, at least 20 hours per week.

If people take meetings, it should only be for a dedicated purpose, and not for social gathering.

Meetings are not social gatherings, so you should carefully consider the people who are to be invited and who are the ones to be merely informed, for example, via meeting minutes.

Dr. Claas Gerding-Reimers

Self-employed / Basler AG

IP Coach / IP Counsel

https://www.linkedin.com/in/dr-claas-gerding-reimers/

CEIPI guest lecturer on IP Management, Top Emerging IP Player (IPR Gorilla Conference), 4x patent applications filed

I help IP'lers and tech business executives grow their business with the right strategy & management in IP & patents.

"The world is what you think it is". (first Huna principle)

Let's connect on LinkedIn and check out www.1stclaasip.de for practical IP management related content, IP Coaching Services and more.

Thank you for reading!

It's been my pleasure that you found the time to read our book.

This book is the first iteration, and its first edition was published in February 2022. It will be continuously improved, corrected, and we will invite other IP experts to contribute.

Are you interested as an IP expert to share your ad-free hacks?

Then please write us an e-mail and get in contact with us at:
contribute@50-ip-hacks.com

Best regards,

Ricardo Cali and the EVOip Team.

Follow Ricardo Cali on LinkedIn and YouTube for more IP insights.

50+ IP HACKS

EVOip – MAKE IDEAS COUNT

Printed in Great Britain
by Amazon

82251320R00068